Nanoscience Research Modules for Pre-Service STEM Teachers

Nanoscience Research Modules for Pre-Service STEM Teachers

Core Nanoscience Concepts as a Vehicle in STEM Education

Clair T. Berube
Hampton University
Shawn T. Dash
Hampton University
Cindy Thomas-Charles
Hampton University

IAP
INFORMATION AGE PUBLISHING, INC.
Charlotte, NC • www.infoagepub.com

Library of Congress Cataloging-in-Publication Data

A CIP record for this book is available from the Library of Congress
http://www.loc.gov

ISBN: 978-1-64113-552-8 (Paperback)
978-1-64113-553-5 (Hardcover)
978-1-64113-554-2 (ebook)

Printed and bound by CPI Group (UK) Ltd, Croydon, CR0 4YY

For Piper, Kennedy, and Collins;
may you always know the power of scientific literacy.
—Clair Berube

To the next generation of teachers and students;
may every learning experience be an adventure. And for Henry
at the beginning of the greatest adventure of them all.

—Shawn Dash

This work is dedicated to the faculty and students
of the School of Science at Hampton University.

—Cindy Thomas-Charles

Contents

Foreword

Elementary, middle and high school science teachers need help. Every day, thousands of people go to work prepared to deliver science instruction to the children of America. Some of these people love science, and some do not. Normally, if you teach secondary science, this means that you are probably a person who likes science and has taken extra science courses in college in order to teach it. Elementary school teachers must teach all subjects during the day, so they are the teachers who are generalists and who may not particularly like science, or prefer it, over the other subjects. Some are even intimidated by it. This book is for them; and for all teachers who endeavor to teach science to their students, whether in elementary, middle or high school.

If you teach science then you know that there are certain topics you prefer to others; life science, physics, or chemistry. But what about nanoscience? Are there any textbooks out there aimed at school teachers that mention nanoscience concepts? This book does. Not only does this book deal with a topic that is hard to find in other textbooks, we have included lesson plans so that teachers can take them and personalize them to fit their learning situations. Nanoscience is an umbrella term under which all of science fits. The world of the very small is the future of science, from experimental to practical applications. This book will show how to teach nanoscience concepts to kids of all ages, which will give them an advantage in the future job market. So whether you love science, studied it in college, or are more of a generalist and are truly terrified, this book is for you. Happy science teaching.

Nanoscience Research Modules for Pre-Service STEM Teachers, page ix

1

What is Nanoscience?

STEM (science, technology, engineering, and mathematics) is a fairly new concept in American education. As separate subjects, science and math have been around for a long time but have rarely been taught as a seamless unit of skills; rather as discrete content areas. This is not how the real world outside of the classroom functions however; in actual research laboratories scientists infuse their science with math, and their math with science, and along with technology and engineering they solve real life problems. In practice you cannot separate the various fields, as you need all of them in order to discover the underpinnings of the natural world, cure a disease, or solve a problem with the space rover. The American future depends on a scientifically literate workforce, armed with knowledge about the laws and theories of science, based on empirical facts instead of beliefs. Only then are we able to rise to the top within global competition. America has competitors who are highly skilled in STEM areas and have the culture and determination to succeed in such technology driven vocations, things that much of the American public does not currently have.

Nanoscience Research Modules for Pre-Service STEM Teachers, pages 1–6

Among American's natural born STEM workforce you will find mostly white, male middle-class people working in every STEM field, from biology to bioethics; quantum physics to space travel; and from cancer research to green technology. There has been a dearth of people of color in STEM for many reasons, including discrimination and cultural ideologies. Why does STEM seem to be only for white men? Where are the African-Americans and women? If STEM jobs are the jobs of the future, how many Americans are going to be left out in the cold because of their lack of education and training in STEM? Some school districts where the poorest minorities live barely have books, let alone enough facilities and staff for the teaching of STEM.

Need for STEM

According to the National Math and Science Initiative (2015)

> By 2018 the economy will have created 46.8 million new jobs. Nearly two thirds of these will require workers with at least some college education, with a slight majority of these requiring workers with a Bachelor's degree or better. The U.S. Census Bureau reported in 2010 only 39% of non-Hispanic whites ages 25–29 had Bachelor's degrees; the numbers were even less for African-Americans (19%) and Hispanics (14%). In addition, there is a shortage of graduates in STEM related disciplines. Economic data show that 1 million additional STEM graduates will be needed over the next decade to fill America's economic demand. STEM based jobs are expected to grow 17% in the next 10 years, outpacing the overall job growth of 10% (p. 1)

All American children are entitled to a proper education; however the United States seems to be missing the mark. In *STEM and the City* (Berube, 2014), the first author discussed how America is falling behind other industrialized countries in the production of scientifically literate citizens:

> According to the National Mathematics Advisory Panel, (2008), approximately 75% of U.S. students are unsuccessful in becoming proficient in math by grade 8 and over 75% are behind by grade 12. In regards to Science, about 10% of students in grade 8 are able to complete advanced science levels on the TIMSS exam, compared to 25% in China and 32% in Singapore (Gonzales, P., Williams, T., Joyelyn, L., Roey, S., Kastberg, D., & S. Brenwald, 2009). While a STEM concentration is moving forward globally, the focus continues to elude many schools nationwide, especially for high-poverty students. (p. 29)

Most of those high-poverty students are African-American. If education is an important component to lifting people out of poverty, then it makes perfect sense that school is the place to do so.

Poverty can only be overcome with knowledge and education. "Poverty, alone, does not determine a student's ability to be successful in these fields. However, studies show that there is a relationship between poverty and the overall likelihood that a student will have access to the kinds of technologies and teachers that can facilitate success in STEM education" (Chauhan & Taylor, 2013, para. 2). The teacher's ability to assist high-poverty students during STEM project-based activities is especially significant. These teachers must demonstrate the ability to actualize STEM activities and use them as a foundation to project-based learning. The critical demand of high-quality teachers and instruction for high-poverty schools is of particular concern for educators nationwide. The research evidence validates the distinctive correlation between student academic achievement and the quality of the educator (Carter, 2001; Haberman, 1995, 2005; Haycock, 2001; Pollard-Durodola, 2003; Rice, 2003). However, the educational community continues to debate what quality and characteristics actually define a quality teacher. Although the *No Child Left Behind* Act of 2002 established criteria for defining highly qualified teachers, the achievement gap and racial disparities continue to dominate high-poverty schools (Peri, Grigg, & Dion, 2005, p. 30).

The quality of the teaching workforce also has a direct impact on the future of America's success in STEM. However, many teachers in impoverished schools have short tenures and the high turnover rate is detrimental to students. Since the United States is experiencing an increasing presence of minority populations, (e.g., African Americans and Latinos) who tend to live in economically restricted areas, students attending such schools critically need an effective teacher. Clearly,

> ... the nation will become more dependent on their contributions to maintain its ascendancy in the global economy. (Clewell & Campbell, 2007, p. 3)
>
> This leads us to the question; can teachers in urban, high-poverty schools provide the necessary quality instruction that address STEM ideals? (Clewell & Campbell, 2007, pp. 29, 30)

Also, how can teachers in urban, high-poverty schools provide a platform for quality instruction and student "buy-in" for STEM?

Nanoscience: The Unifying Concept in STEM

One of the groups responsible for fixing this problem should be schools and colleges of education across the country; places where teachers are educated. As a general rule, most teachers are uncomfortable teaching science content. According to Ward (2008),

> The reality of the situation is that the majority of in-service elementary teachers have only superficial understanding of science concepts at best and often the same misconception as their students. (p. 2)
>
> In terms of science teaching, most elementary teachers avoid teaching science or resort to coping strategies, such as textbook worksheet styles of teaching, with occasional "science- related" activities, featuring more process related aspects of science rather than conceptual aspects, and avoiding questioning and discussion. (p. 2)

Adding to that is the fact that physics and the world of quanta (nano) is the most intimidating of all, yet nano is the cutting-edge future of science. If teachers across America were trained with fundamental and impending scientific concepts in their science-methods courses at the university level, scientific literacy can only dramatically improve. Nanoscience is one such concept; as it is multidisciplinary in nature and is regarded as the basis for innovated technologies in many fields (Sohlberg, 2006). Introducing Nanoscience into lesson plans in science-methods classes is vital. At the elementary and secondary level, teachers who instruct in science must be scientifically literate in these topics in order to prepare children for possible careers in STEM, especially for students in high-poverty schools. This training can only come from teacher education programs in American colleges and universities.

Scope and Purpose

The authors of this book seek to provide pre-service science teachers with high-quality nanoscience STEM modules, with which to create lesson plans and problem-based lessons to use in their future classrooms, both at the elementary and secondary level. In order to do so, the authors received a grant from the National Science Foundation entitled "HBCU-UP ACE NANO HU: Infusing Nanotechnology into Science Education Methods Courses for Pre-Service Teachers at Hampton University." Hampton University is an historically Black college founded in 1868 for the education of freed slaves and Native Americans. Dr. Dash is an Assistant Professor of Evolutionary Biology and Dr. Berube is an Associate Professor of Science Education.

Nanoscience was chosen since its applications reaches across virtually every scientific field; from biology to physics for that matter all STEM domains. Nanometer refers to a unit one billionth of a meter, which is 0.000000001 of a meter (10,000 smaller than a human hair). To provide some perspective a human hair is 17,000 nm thick, a #2 pencil is 190,000,000 nanometers long and President Obama is 1,850,000,000 nm tall. Nanoscience has discovered that when studying materials at such scales, the properties of materials change, a common example of this is gold. At the macroscopic scale (millimeter on-up) gold appears gold in coloration however, when examining gold nanoparticles these groupings appear reddish; with other unique properties. Nanoscience is focused on discovering and understanding how physical, chemical, and biological processes behave at the nanoscale.

The core concepts of nanoscience (modified from Light et al., 2007) can be summarized as follows:

- Size and Scale
- Size Dependent Properties
- Particulate Nature of Matter
- Tools and Instrumentation/Characterization Modeling
- Quantum Mechanics
- Dominant Forces
- Surface Area to Volume Ratio
- Self-Assembly
- Surface Dominated Behavior
- Technology and Society; and
- Public Education

Though incorrectly used interchangeably, nanotechnology uses nanoscience as a means to manipulate matter on an atomic, molecular, and supramolecular (describing structures composed of several molecules) scale. Each of the various core concepts allow generation of activities that fit into school curricula and encompass STEM fields.

The Scanning Tunneling Electron Microscope, invented in 1981 by IBM, allows us to "see" single atoms by scanning the surface of them with an extremely small, fine needle, where the electric voltage at the tip reacts with the electron on the metal surface, thus producing an image (Bowles, 2004, p. 3). Every teacher and student has been in contact with nanoproducts, that is anything that is made with or enhanced by nanotechnology. The proliferation of these products are in all aspect of life, from make-up, sunscreen (using zinc oxide nanoparticles), antimicrobial socks (using silver nanoparticles), waterproof materials (nanoparticles producing super hydrophobic coatings), time-delayed medicines, electronics, and even in the food industry (Sondi & Salopek-Sondi, 2004; Filipponi, Sutherland, & Center, 2010).

Chang (2006) explores how Nanoscience can spark the interest of students in school so as to make science more appealing. He states:

> Kids are curious by nature, yet students rarely have opportunities to satisfy their curiosities in today's classrooms. The latest scientific discoveries and cutting-edge technologies are often left off classroom menus and too many barriers tend to separate those who "do" science from those who teach it. Perhaps most importantly, science remains divided into disciplines like biology, chemistry and physics, and is usually taught separately from math and engineering. Many of us believe these arbitrary disciplinary divisions are responsible for the worsening crisis of scientific disinterest among our young people. (p. 6)

Since Nanoscience covers every scientific discipline, it is the unifying component in STEM, and one that needs to be taught in every science classroom in America. Indeed, Nanoscience technology is the interdisciplinary link to all of STEM.

> It is becoming clear that interdisciplinary is critical to solving global challenges. United Nations Educational, Scientific and Cultural Organization (UNESCO), for instance, has established "Integrated Science Education" as a major program area. Nanotechnology research is at an all-time high and "nano" literacy has become a global priority. Roughly two million new nanoscientists and engineers will be needed worldwide in the next 10–15 years. (Chang, 2006, pp. 6, 7)

This book seeks to ensure that nanoscience is taught in American classrooms in order to achieve this goal.

2

The Background Study

Methodology

NanoHU: Nanoscience Transforming STEM Education at Hampton University

This book is a result of a two-phase research initiative, Phase 1 (creating nano modules—lesson plans—with high-school students at NanoHU Pioneers Summer institute with Drs. Dash and Berube as scientist mentors) and Phase 2 (incorporating nano modules into science methods courses at the university level via Dr. Berube's science methods courses). Science methods courses are taught in colleges and universities across the county by professors who prepare pre-service teachers; or college students preparing to become either elementary or secondary teachers and science teachers. Science methods courses typically combine science content with pedagogy; students learn science and also how to teach it. We had two specific aims for this project and they include (a) the facilitation of student-driven research (high-school students selected for the study) and (b) the application of that research to evaluate and improve nanoscience education for primary and secondary school teachers thereby reaching K–12 students.

Nanoscience Research Modules for Pre-Service STEM Teachers, pages 7–11

Phase 1

Phase 1 focused on a nanoscience summer STEM institute (NanoHU Pioneers Summer Researchers) where selected local high-school students were charged with conducting research with university professors. Included in the nanoscience summer STEM institute at Hampton University were several STEM faculty and local area high-school students chosen especially for this program. This book involves the research of Drs. Dash and Berube and three high-school student mentees that were assigned with the development of nanoscience learning modules, or lesson plans. These modules consist of nanoscience lesson plans that the high-school students created and experimented with during the summer institute. The ultimate goal of these lesson plans would be for middle or high-school teachers to introduce and teach nanoscience lessons to their students.

The NanoHU institute was held at Hampton University across all STEM departments. A total of 16 students were partnered with various STEM professors; Drs. Dash and Berube worked with three high school students in the biology department. During the summer of 2015 of Hampton University's 2015–2016 Academic year, nanoscience high-school student mentees met for 6 weeks at the University for the student driven work. Students were accountable for the development, extension, and delivery of nanoscience demonstrations and experiments to be utilized later in nanoscience education. The nanoscience students developed as researchers with hands-on research, skills training, literature review competences and enhanced their science communication skills. All mentees were integrated with the HU-Nanoscience program with the expectation that the students would be active participants in nanoscience-related seminars, courses, and internships.

High School Participants:

1. Kevon Barrett a 10th grader at Phoebus High School in Hampton Virginia.
2. Sydney Hardy a 10th grader at the Governors STEM Academy at Grassfield High School in Chesapeake Virginia.
3. Tomar Pettus Jr. an 11th grader at Tabb High School in Yorktown Virginia.

Phase 2

Additionally, this project sought to incorporate nanoscience technology and problem-based learning into science methods courses at Hampton

University. Problem-based learning (PBL) has been a research area for Dr. Berube and studies show that pre-service STEM teachers gain more self-confidence and self-efficacy if they are taught how to conduct PBL STEM lessons in their classes (Thomas et al., 2013). The goal is for college and university science and STEM education professors to teach these nanoscience modules (lesson plans) to pre-service elementary teachers with problem-based learning lessons (PBLs) as a means to expose and engage STEM educators via science methods courses at the university level. Dr. Berube teaches elementary and secondary science methods courses and teaches Nano modules in her classes.

Fundamental to pre-service elementary and secondary teachers is being well-prepared as scientifically literate educators to ensure their students are equipped for future employment in STEM fields. Problem Based Learning lessons (PBLs) are simply science lessons that are taught via a real life problem presented to the students, where the students have to become scientists in order to solve the problem.

Problem Based Learning (PBL) as a Means to Teach Nanoscience to Pre-Service Teachers

What is PBL?

Virginia Initiative for Science Teaching and Achievement (VISTA) researchers define problem-based learning (PBL) as "students solving a complex problem with multiple solutions over time like a scientist in a real-world-context" (VISTA, 2015). They further state the problem must be meaningful to students and is typically embedded in a course of study one to five weeks in duration (VISTA, 2015). Through PBL students ask scientific questions relevant to their lives, collect evidence, and develop explanations based on the evidence obtained. This type of inquiry provides students with the highest level of investigative control, unlike traditional teacher-led explorations (Bell, Smetana, & Binns, 2005). Students use The Problem-Solving Cycle as a roadmap throughout their PBL investigations. Contrary to the lockstep myth of The Scientific Method, The Problem-Solving Cycle allows students the flexibility to move forward or retrace their steps in the investigation as needed. This enables student researchers to backtrack in response to new information gained and better represents the way scientists work to find solutions in their profession. PBL has been shown to have benefits to both students and teachers, including increased time on task (Drake & Long, 2009), increased professional confidence (Etherington,

2003), and improved student behavior (Reeves, 2006; Skinner & Chi, 2012; Thomas et al., 2013).

Phase 1 Timeline and Date Management

Over the 6-week summer research session, students proactively participated in hands-on research projects developing research and science communication skills.

May	June	July	August
Recruit Students			
	Student Project Development		
	Student Nanoscience Review		
	Student Experimentation & Implementation		
		Composition of Manual	
			Post-Project Development of Teacher Resources

Our timetable is divided into three phases to effectively manage students and research directives, with: (a) preparations for the overall projects with nanoHU students, (b) carrying-out research for both realms of the project including the nanoHU students and preparation of resources to be used by preservice teachers in science methods courses, and (c) summations of findings. The Phase 1 summer session included student (high-school) development of a nanoscience project, funded by an NSF grant and taking place at Hampton University (NanoHU Program).

Weeks 1–2

Students became familiar with research techniques, basic concepts of nanoscience within the context of this study. Drs. Dash and Berube worked with the three high school student researchers in the department of biology.

Weeks 2–6

NanoHU Students completed their literature review on nanoscience education. Working within the NanoHU program in the department of biology, students cultivated skill-sets by developing, experimenting with, and improving projects that influenced the learning modules based on the selection of core nanoscience concepts. Students also generated their own novel demonstrations and activities.

Dr. Dash set various weekly aims and goals for NanoHU student researchers that were discussed during laboratory and skill building sessions. During this time students provided updates, troubleshooting solutions, data collection, and performed group brainstorming sessions. Time was devoted to proper student supervision and mentoring. With all data collected and comprising a weekly reported provided to HU nanoscience program.

After completion of Week 2, all students in the nanoHU program, including the three students mentored by Drs. Dash and Berube, presented PowerPoint presentations explaining their particular area of research concentration being conducted with their mentors. Dr. Dash and Berube's students reported to the larger group the research and review of literature on nanoscience and how the team was going to use nanoscience to create learning modules for the use of pre-service STEM teachers in science methods courses at Hampton University. The nanoHU summer researchers compiled and evaluated a number of preexisting educational guides (http:// nanoyou.eu/ is a consortium of European Schools teaching nanotechnology. The site has many useful activities and demos as well as an educators section. The NISE Network http://www.nisenet.org/ a national community of researchers and informal science educators dedicated to fostering public awareness, engagement, and understanding of nanoscale science, engineering, and technology. As well as the National Nanotechnology Infrastructure Network of education and training portal http://www.nnin.org/education-training/k-12-teachers/nanotechnology-curriculum-materials) for activities and experiments that would fit a set of student-generated criteria: how long does the experiment take to complete (so pre-service teachers can complete it), how easy is the activity to understand, does the experiment have real-world application, does it provide some sense of a hands-on experience, and does it teach a core nanoscience concept?

3

The Manual

Nanoscience Manual

The manual and modules were products of student research, experimentation, and manipulation and therefore represented a student perspective on resources, explanations, and "what is fun." Within the manual are seven modules:

Module 1. Metric Units and the Nanoscale
Module 2. Nanoscience and Technology
Module 3. Nanoproducts
Module 4. Silver nanoparticles as an Antimicrobial
Module 5. Producing Nanoparticles
Module 6. Nano-encapsulation
Module 7. Nanoscience in nature: Build a Beetle

These modules were design to be student-based learning activities where students would be required to ask questions or generate hypotheses before the activity and re-evaluate their ideas after the module was completed. As previously noted the nanoHU summer researchers attempted to include core

Nanoscience Research Modules for Pre-Service STEM Teachers, pages 13–15

concepts from nanoscience. The concepts of Size and Scale were addressed in Modules 1, 3, and 5. Size Dependent Properties were discussed in Modules 3, 4, 5, and 6. Particulate Nature of Matter was expanded on Modules 4 and 5, Tools and Instrumentation were explored in Module 5, Surface Area to Volume Ratio was addressed in Module 6. The concepts of Technology and Society were the focus of Modules 2 and 3. Finally we can argue all modules meet the concept of Public Education, especially in Module 7 where students watch a film on desert beetles, see nanoscience in nature, and create their own hydrophobic beetle using bio-inspired (biomimicry) nanotechnology.

Phase 2

Fall 2015

Dr. Berube's classes worked through problem-based learning modules with pre-service elementary and secondary science and mathematics teachers in a science methods course during the Fall 2015 academic semester. Dr. Dash guest lectured at the preliminary introduction to the concept of nanoscience.

Spring 2016

Analysis of problem based learning activities. Results of the project were used to augment and refine the pedagogical approaches to develop best practices for nanoscience education. While materials typically used in demonstrations and educational activities are developed by nanoscience, this project is unique in that high school students represented a filter and evaluation mechanism to develop and maintain a certain level of engagement with contemporary secondary students. The high school researchers were a valuable resource in determining which modules would be of interest to other students in school systems like themselves, in order to attract elementary and secondary students to nanoscience.

September	October	November	December
Assessment of Nanostudent Project			
Pre-assessment			
Introduction of Pre-Service Teachers to Nanoscience Experiments			
	Presentation of Nanoscience Experiments in a Problem Based Setting		
			Post-assessment

The following manual was created with the goal that college professors across America would be inspired to use this book and books like it in order to teach their college students (future teachers of America) how to incorporate nanoscience into their STEM lessons. STEM subjects can be intimidating to any teacher, and this project will hopefully ensure that American students will grow up learning about nanoscience so that they can follow their STEM career goals as they go on to college or beyond.

Nanoscience in the Classroom

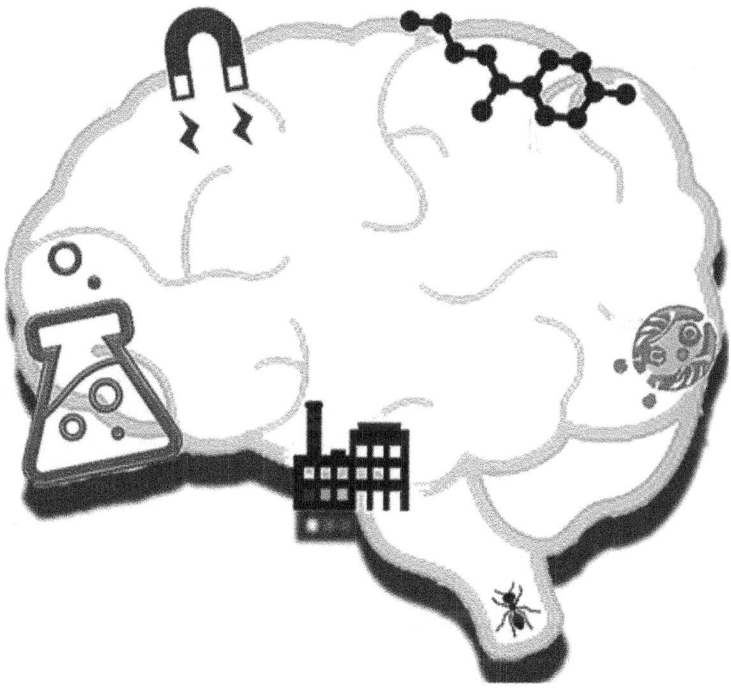

Nanoscience Module Set for Pre-Service Teachers

NANOHU PIONEER SUMMER 2015 RESEARCH

HAMPTON
UNIVERSITY

Hampton University is a dynamic, progressive institution of higher education, providing a broad range of technical, liberal arts, and graduate degree programs. In addition to being one of the top historically black universities in the world, Hampton University is A tightly knit community of learners and educators.

NanoHU
NANOSCIENCE
TRANSFORMING STEM EDUCATION
AT
HAMPTON UNIVERSITY

NanoHU is a newly developed program at Hampton University that affords students to engage in cutting-edge opportunities associated with nanoscience. NanoHU is a cooperative effort between the departments of Biological Sciences, Computer Science, Chemistry, Engineering, Mathematics, Marine & Environmental Science, and Physics at Hampton University.

NSF

This project as made possible by a sub-award of The National Science Foundation Grant NSF 12-519, Historically Black Colleges and Universities Undergraduate Program (HBCU-UP).

This book had contributions by NanoHU Pioneer Summer Research Students. It represents their six weeks of hands-on research and dedication. The students had two mentors, Drs. Dash and Berube, who guided the students in their explorations, wrote, compiled, and edited the final product.

Table of Contents

Introduction

An emerging area within science, technology, engineering, and mathematics (STEM) fields is that of nanosciences. This interdisciplinary science spans all aspects of technology and engineering, manufacturing, biology, chemistry, physics, economics, policy, and so on. Though STEM accomplishments are continuing to expand, STEM education in the United States is argued to be not only at a standstill but seriously falling behind (Gonzales, Williams, Joyelyn, Roey, Kastberg, & Brenwald, 2009). Two major causes cited for STEM abandonment are: (a) limited qualified STEM educators, and (b) a serious lack of students engaged in STEM subjects (NMSI). A continual challenge of STEM education for at least the last decade is how to excite as well as retain students in science and related fields (Nichols et al., 2007). Nanoscience is a field that is relatively young and presents itself as exciting with little effort to convince students to be interested (Sohlberg, 2006). We view nanoscience as the gateway treatment to increase interest in STEM given nanoscience's novel appeal and integrative network.

A directed effort is needed for United States to produce a citizenry that is scientifically literate. One area that does not only provide the necessary theory of science but also the broad and *practical* application is nanoscience. Nanoscientists focus on discovering and understanding physical, chemical, and biological processes which occur at an atomic or molecular scale, that is the nanoscale. A nanometer being one billionth of a meter, or three and half gold atoms wide (Filipponi & Sutherland, 2010), or in another perspective the width of a U.S. dime is 17,900, 000 nanometers. Related is the more widely used term nanotechnology which is defined as the application, manipulation and design of materials at the nanoscale. A handicap in conveying STEM concepts and also in long-term recruitment is that grade school setting learning activities do not provide a practical context to students and may produce poor experiences in the classroom (Nicholls et al., 2007; Price, 2012; Smith et al., 2008). Nanoscience is a field that is relatively young and presents itself as exciting with little need to convince students to be interested (Sohlberg, 2006). In fact nanotechnology is reshaping the way we build our world, and providing a practical foundation to lure students to science; with nano-circuits, nanofibers, water repellant nano-coatings, pharmaceutical delivery, antimicrobial films, and as a platform itself for education.

The core concepts of nanoscience (modified from Light et al., 2007) can be summarized as follows:

> Size and Scale, Size Dependent Properties, Particulate Nature of Matter, Tools and Instrumentation/Characterization Modeling, Quantum Mechanics, Dominant Forces, Surface Area to Volume Ratio, Self-Assembly, Surface Dominated Behavior, Technology and Society, and Public Education.

The education and cognitive sciences agree that students learn best when they are engaged via: experimentation, in-depth investigations, inquiry-based learning, and practical activities; and when such activities are supplemented by approaches that demonstrate and provide a practical and applicable context (Krajcik et al., 1998; Michael, 2006; Summers & Hrabowski, 2006; Wood 2013).

NanoHU Pioneers Summer Research Program

The high-school STEM students chosen to work with Drs. Dash and Berube were charged with the contributions towards production of a manual to introduce pre-service teachers to nanoscience and provide a context to teaching nanoscience by addressing its challenges and rewards. As one reviews the numerous and varied experiments and demonstrations for nanoscience one thing is apparent: many require a lot of set-up and time to conduct. As this manual is intended for pre-service teachers, with limited space, time, and supplies, we have spent effort to evaluate previous activities for ease of set-up and execution.

The manual presented is a series of learning modules focused on core concepts in nanoscience (Nanoscale, Size Dependent Properties: differences at the nanoscale, and, Surface Area to Volume Ratio: encapsulation, producing nanoparticles, Technology and Society: application of nanoparticles as an antimicrobial and exploring nano-products). Though the modules are typical introductory nanoscience activities, the spirit of the guide is important to note. Hampton University nano-students have evaluated these classic exercises for modes of engagement, skills learned, and efficiency of conveying nanoscience, making numerous critical changes to previous designs. In addition to addressing aspects of instruction another goal was the hope that one activity would be a "spark" project that would excite students to become more deeply involved in STEM.

Pedagogy: Teaching Nanoscience in Elementary and Secondary Science Classrooms

For science teachers across America who are trying to make a difference to their students, it is always important to try to make learning relevant to their lives. What does this look like? How do we as teachers make connections for our students between the science classroom and the greater world? Expanding content and pedagogical practices to engage their curiosities and to prepare them for the future job markets are all crucial. Nanoscience is one of the main overarching themes in every scientific field and has the potential to transform the world as we know it. Indeed, it has already changed the planet; from computer chips to nanobots to medicine, the field of nanoscience is extremely relevant to modern society. American science teachers at all K–12 levels can take advantage of such a rich content area to make their classrooms real places of learning and growth.

Problem-based learning is the perfect vehicle for the introduction of nanoscience into science classrooms and labs. As mentioned earlier, PBL (problem-based learning or problem-based lessons) are an exciting way to engage students in high-level cognitive skills. Imbedding a lesson in a real-life problem and then allowing the students to solve that problem has shown to raise students' science self-efficacy and confidence (Thomas et al., 2013).

Background and Resources for Educators

- http://nanoyou.eu/ is a consortium of European Schools teaching nanotechonology. Site has many useful activities and demos as well as an educators section.
- http://www.nisenet.org/ The NISE Network is a national community of researchers and informal science educators dedicated to fostering public awareness, engagement, and understanding of nanoscale science, engineering, and technology.
- http://www.nnin.org/education-training/k-12-teachers/nanotechnology-curriculum-materials
- Nanotechnology Classroom Activities and Curriculum Materials developed by the National Nanotechnology Infrastructure Network.

- http://www.nanoandme.org/home/ Our aim with nano and me is to provide balanced information about nanotechnologies and be the hub of debate for everyone to discuss the important issues which arise from its use.
- http://www.nnin.org/sites/default/files/files/NNIN_Outreach_Demo_Guide-11_10_0.pdf National Nanotechnology Infrastructure Network Outreach Demo Guide
- http://www.nano.gov/education-training/teacher-resources Site devoted to nanotechnology education.
- http://www.tntg.org/documents/Teaching%20Nanotechnology%20in%20the%20High%20School%20Curriculum.pdf *Teaching Nanotechnology in the High School Curriculum: A Teacher's Guide.*

All figures were produced by S. T. Dash (except the candies for Module and materials).

1

Module 1: Metric Units and the Nanoscale

Background

One of the harder concepts to covey to students is the idea of size and scale. Where we focus on using the metric system to determine size and scale to allow us to communicate in terms to convey relationship to size; we say an object is tiny but what is tiny? Is it simply that millimeters are used to measure tiny? What do we call something that uses micrometers to record the size? In this module students will be engaged in a hands-on activity that will provide a basis for student understanding of metric measurements and viewing the world at both macro and microscale. These activities will introduce students to a scale of size, by using relative sizes of various objects placed along a rope representing size and scale with distances, as well as asking students to match novel objects of unknown size to correct size/scale position applying what lessons they learned in earlier activities in a game-like setting.

Background for Educators

Informative PowerPoint Summarizing of nanoscience

http://nanoyou.eu/attachments/298_Power%20point%201%20Discover%20the%20nanoworld.pdf

Definition of a meter

http://www.bipm.org/jsp/en/ViewCGPMResolution.jsp?CGPM=17&RES=1

Student Learning Objectives

SLO 1 Students will be able to determine scale of metric units.

Example: nanometer 10^{-9}

SLO 2 Students will be able to correctly place objects in positions of relative size.

SLO 3 Students will be able to correctly apply size and scale to novel objects.

Materials and Methods

Materials

1. (1) 30.5 meter rope (100 foot). A larger rope can be used if facilities permit, the key is to have a long enough rope that coveys distance (scale).
2. Different sets of note cards, 1 colored set around 20 note cards, and one larger set of white cards (100 cards).
   a. Color note cards will indicate unit of measure, and size see chart below.
   b. Plain note cards will report various objects there should be around 30–40 objects that can be used for student teams to place during the game portion.
3. Binder clips
4. Sharpie
5. Objects that represent items of various size, for example: human (action figure), red blood cell (microscope slide) you can use a red plate as a stand in, pollen (microscope slide) you can use fake flowers as a stand in, hair (width), virus (model), mobile phone (iPhone) cellular device, dog (plush toy), bug (plastic models).

Methods

Students are (re)introduced to the metric system, we will focus on the system and microscale in the box below.

base unit		10^0		1
deci	d	10^{-1}	1/10	0.1
centi	c	10^{-2}	1/100	0.01
milli	m	10^{-3}	1/1,000	0.001
micro	µ	10^{-6}	1/1,000,000	0.000001
nano	n	10^{-9}	1/1,000,000,000	0.000000001

1. Students are given the rope, and asked how they would divide the rope into correct units and at a proper scale to represent m, dm, cm, mm, μm, and nm. (One end of the rope will represent meters the other end will represent nanometers.) Allow students time to work through this problem.

   In the end, decimeters will occur at 3 m (10 ft.) as dm is 10^{-1}, therefore cm will occur at 6 m (20 ft.) = 10^{-2}, and so on. Students should easily see that there is a large distance between the meter and nanometer. Using a colored note card write m 10^{0} at the dm use a colored note card and record dm 10^{-1} these cards will act as flags and place holders for all our units, use the binder clips to hold in place. Ask students for their thoughts on the difference between the two. Does this covey the difference in scale?
2. Give students various objects. Students should be encouraged to explore these objects and determine the size of each. Discuss with the students the appropriate unit of measurement (when should we use meters or micrometers). If students cannot determine the size they may research the sizes of various objects.
3. The students will now place the objects on the correct region of the roped based on their measurements. The action figure at the meter site, the red blood cell-plate will be placed near the micrometer position and the virus at the nanometer end for example.

Figure 3.1 Nano-rope up close. Students having placed the flowers representing pollen near the micrometer portion of metric system rope.

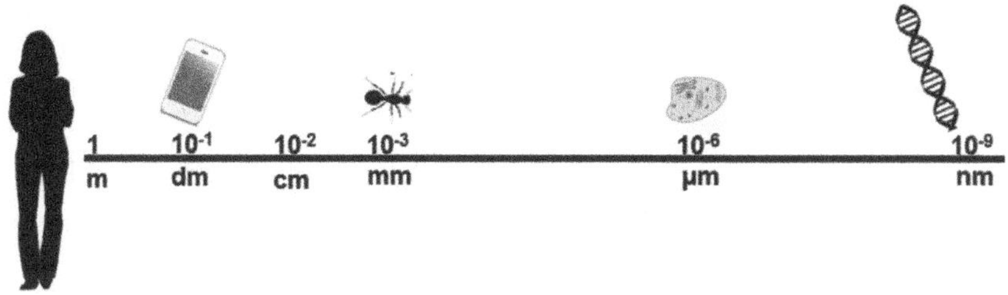

Figure 3.2 In the end the set-up should be similar to this image.

The teacher follows this placement exercise with a summary of what is the context thus far, what are the relationships to the world we see m-mm compared to µm and nm. Introduce some points that reference nanoscience and the nanoscale. Check if this exercise allowed the students to get a better understanding.

Figure 3.3 Nanorope of Science. The left image represents the meter end of the rope, while the student with the red ball (red blood cell) represents the µm and the far students is at the nanometer end. The right image represents the nanometer looking back towards the meter.

4. The student teams are given note cards with the names of various everyday objects written on them. They will then race to place those cards in the correct region at which those objects can be measured and observed.
5. The teacher will then evaluate card placement and see which person or group was the winning team.

Class Questions

1. At what scale do most people view the world?
2. What are the relationships between various units in the metric scale?
3. What is the nanoscale?
4. What unit should we use to measure human bodies, human cells, human molecules?
5. By knowing about the nanoscale, how do you think it has affected our lives?
6. Why is being familiar with the nanoscale important?

References and Suggested Materials

National Nanotechnology Infrastructure Network Outreach Demonstration Guide

A wonderful guide containing many demonstrations that can be used to expand on nanoscience. http://www.nnin.org/sites/default/files/files/NNIN_Outreach_Demo_Guide-11_10_0.pdf

What's the Smallest Thing You Know?

This lesson is to encourage students to think about the size of objects through the use of literature and mathematics. http://www.nnin.org/sites/default/files/files/What%27s%20the%20Smallest%20Thing%20You%20Know_TG_1.pdf

The Metric System from Big to Small

This lesson will review the metric system and relate the size and scale of objects to various metric units, including the very small nano-scale. http://www.nnin.org/sites/default/files/files/NNIN-1168_0.pdf

2

Module 2: Nanoscience and Technology

Background

Frequently in science there are ideas that are similar; for example nanoscience and nanotechnology. Nanoscience is defined by the Texas Materials Institute as the study of structures and materials on the scale of nanometers. In contrast nanotechnology uses nanoscience as a means to manipulate matter on an atomic, molecular, and supramolecular scale. A great place for an overview is *What is Nanotechnology* (http://www.nano.gov/nanotech-101/what/definition). Using class discussion and team think-pair-share activities, students will explore and develop a general knowledge of nanoscience/nanotechnology. In addition, students will be able to list a range of industries where nanotechnology is applied.

Background for Educators

Nanowerk Nanotechnology Neat Stuff
Amazing Nanotechnology Links, Images, and Sites for Kids
http://www.nanowerk.com/n_neatstuff.php

Texas Materials Institute
http://tmi.utexas.edu/resources/what-is-nanoscience/

Science, Engineering, and Technology
Distinct but interconnected discussion on the relationships of science, technology and engineering.
http://www.eie.org/overview/science-engineering-technology

Student Learning Objectives

SLO 1 Students will be able to compare and contrast science and technology.

SLO 2 Students will be able to generate a Venn Diagram to compare nanoscience and nanotechnology.

SLO 3 Students will be able to Draw connections between nanoscience and STEM themes.

Materials and Methods

Materials

1. Large notepad—Giant Post-its
2. Colored markers

Methods

1. Have students in groups create a list of ideas, concepts, impressions, and thoughts about what is the difference between science and technology.
   a. Assign a team recorder and determine a set number of ideas for comparison and contrast that each member needs to provide.
2. Collect all student lists and provide a context of the similarities and differences between science and technology. A great site for an overview as noted above is *Science, Engineering, and Technology*. Take some time by using the following devices (Figure 3.4 and Table 3.2) to instill an understanding of the two fields.
3. Students should be given other groups' constructed comparisons to examine for similarities to their own ideas as well as new sights from the lesson and peer's thoughts.

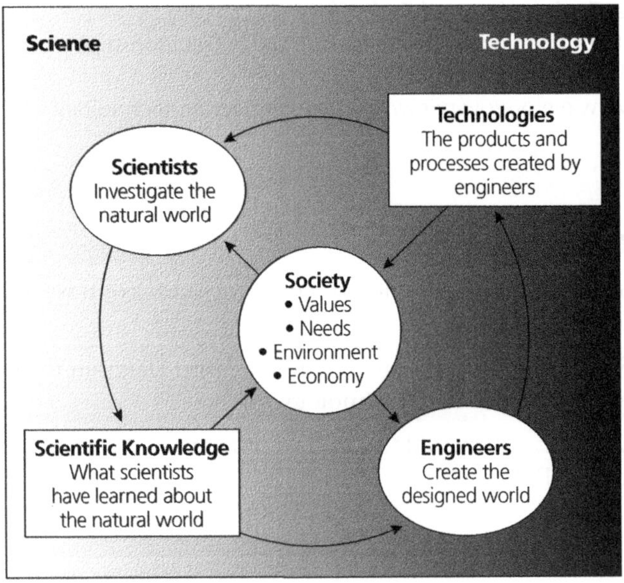

Figure 3.4 Relationships between Science and Technology. Taken from http://www.eie.org/overview/science-engineering-technology

TABLE 3.2 The Relationships of Science to Technology to Engineering

Science	Engineering	Technology
...is the body of knowledge of the physical and natural worlds	...is the application of knowledge in order to design, build, and maintain technologies	...is the body of knowledge, systems, processes, and artifacts that results from engineering
...seeks to describe and understand the natural world and its physical properties	...seeks solutions for societal problems, and needs, and wants	...can be used to describe almost anything made by humans to solve a problem or meet a need
...uses varied approaches—scientific methods such as controlled experiments or longitudinal observational studies—to generate knowledge	...uses varied approaches—for example, engineering design processes or engineering analyses—to produce and evaluate solutions and technologies	...results from the process of engineering
Scientific knowledge can be used to make predictions	Engineering aims to produce the best solutions given resources and constraints	Technologies are anything made by humans to fill a need or desire

4. Have students view as either part of homework or during class:
   - A Slide introduction to Nanoscience (very basic) https://www.youtube.com/watch?v=7tqUvxR20_I
   - What is Nanotechnology. This Video Will Blow Your Mind https://www.youtube.com/watch?v=XZVH7CASrBk
5. Ask students to generate a Venn diagram on a large post-it pad that compares nanoscience and nanotechnology. Post the Venn Diagrams.
6. Have one student become the group spokesperson and stand by each group's diagram. Allow all students to interact and examine each other's work asking questions regarding the construction of their diagrams and ideas. Half-way through this tour of diagrams ask each group to select a new spokesperson to switch so they may continue the tour.
7. After all students have completed their tour, recall the groups and pose the idea that nanoscience is deeply related to STEM fields. Ask the class to provide their ideas of how nanoscience is related to science, technology, engineering, and math.

Class Questions

1. Do you think your answers would have been different if you did not discuss science and technology first?
2. Do other groups ideas influence the way you thought about the Venn Diagram?
3. What were some repeated ideas through each of the three discussion sets?
4. Do you think your ideas will change as you are exposed to more nanoscience activities?

References and Suggested Materials

Light, G., Swarat, S., Park, E. J., Drane, D., Tevaarwerk, E., & Mason, T. (2007). Understanding undergraduate students' conceptions of a core nanoscience concept: Size and scale. In *Proceedings of the First International Conference on Research in Engineering Education,* (pp. 1–13).

Winkelmann, K., Bernas, L., & Saleh, M. (2014). A review of nanotechnology learning resources for K–12, college and informal educators. *Journal of Nano Education, 6*(1), 1–11.

3

Module 3: Nanoproducts

Background

Anything that is made with or enhanced by nanotechnology is considered to be a nanoproduct. The purpose of this module is to introduce nano products. In the past few years the production of nano enhanced products has increased. Mainly because the knowledge of "nano" has increased. Some of the most popular nanoproducts include: sunscreen (zinc oxide), makeup (Zinc Oxide {OPTIONAL}), and EverDry. As you can probably tell, these items are not that complex and they apply to our everyday lives. For example, without zinc oxide the formulation for topical sunscreens would be limited and you would risk exposure to harmful radiation from the sun. Without makeup you could not cover up tiny scars and blemishes before a big day. The most recent would be EverDry. When applied to anything, this EverDry concoction gives whatever it's applied to a hydrophobic nanocoating and can keep things from getting wet or stained. Those are just examples.

Background for Educators

Amazing Nanogold, Why is it red?
https://www.youtube.com/watch?v=MsnfvNd6qTc

Gold Nanoparticle Demo
Student Learning Objectives

SLO 1 Students will be able to see the different properties of nanoproducts.

SLO 2 Students will be able to comprehend and apply the concepts of nanoscience.

Pre-Lab Questions

1. Given your current knowledge, what do you think nanoproducts are?
2. Where in everyday life do you think nanoparticles would be the most beneficial
3. How can you tell if a product has nanomaterials in it?
4. Where do you think nanoscience can be the most dangerous in terms of the environment and human life?

NanoGold Demo

Background

At the nanoscale things don't always behave as they do when they are larger. Nanoparticles may be stronger or lighter, or, more reactive because of their small size. As such, nanoparticles may be used in a variety of ways that the larger forms cannot be used.

Student Learning Objectives

SLO 1 Students will be able to make observations and supply potential explanations of observations based on previous knowledge to novel situation.

SLO 2 Students can create discussions on how adding nanomaterials to normal objects can change the entire composition of the objects in question.

Materials and Methods

Materials

- 4–6 (depending on class size) 1 dram screw tops vials filled with Meso-Gold (https://www.purestcolloids.com)
- 4–6 (depending on class size) 1 dram screw tops vials filled with gold leaves or Imitation gold leaf (from art supply shop) we just need something that looks like macroscale gold (Figure 3.5).

Methods

- Pass vials around the room so students may examine the contents.
- Request students to determine what they think are in each vial, and discuss their thoughts.
- Inform students that the same materials at the nanoscale have different properties and behave differently. Here pass around a vial of gold leaf (macro-scale) and then a vial of gold nanoparticles (nano-scale) again asking the students if they think that both are gold. Inform the students that yes both are in fact gold but the red-liquid is gold nanoparticles measuring 3.2 nm.

Figure 3.5 Vials of Gold. (a) displaying gold leaf and typical image of gold, (b) gold nanoparticles

- **Ask students, what sample has larger atoms?** This is a great chance to explore the perceptions of size of an object related to size of atoms comprising the object. Remind students that 1 nanometer is only about 3.5 gold atoms wide.
- **Ask students, why there may be a difference in color?** This is a good way of getting students to start talking about how nanomaterials can have different properties compared to macro materials.
- **Ask students, whether or not they think that in addition to being a different color, gold nanoparticles would have other properties that differ.**

Nanosand Demo

Background

Sand is a natural occurring hydrophilic material composed of finely divided rock and mineral particles. It is defined by size as being finer than gravel and coarser than salt, all of which are hydrophilic naturally. Sand can also refer to a textural class of soil or soil type. It is also filled with pores that allow the sand to absorb water. The absorbency of the sand depends on the texture of the grains.

Figure 3.6 Nanosand smiley face. Nanosand shaped underwater to make a smiley face, clearly visible are the air pockets around the sand that keep the sand unexposed to water while submerged.

Hydrophobic is the physical property of a molecule that repel the uneven polarity of water (Is Salt Hydrophobic?). Water is a polar molecule, meaning it has uneven charges on its respective molecules. As such, hydrophobic molecules tend to be non-polar. Some hydrophobic molecules include oils and fats. Scientists have found ways to emulate the hydrophobic properties of non-polar materials and create synthetic water-repellant coatings, such as nanosand, as a way to protect water-sensitive goods without spending extra on high-end sealers.

Nanosand allows for the demonstration of how changing certain properties in the nanoscale can affect how that material behaves at the macroscale. Recall that hydrophobic surfaces repel water; what allows nanosand to repel water despite not being hydrophobic is a hydrophobic coating comprised of hydrophilic nanoparticles on the grains of sand. You can compare the difference of how nanosand and regular sand interact with water and see if the hydrophobic coating of the nanosand makes any noticeable differences.

Background for Educators

Magic Sand

www.nisenet.org/catalog/programs/magic_sand_nanosurfaces

Student Learning Objectives

SLO 1 Students will be able to understand the difference between nanoscience and the rest of the world.

SLO 2 Students can create discussions on the application of nanomaterials.

SLO 3 Students will make observations on the nanoscale.

Pre-Lab Questions

1. How is sand created?
2. How dose sand react when in contact with water?
3. What does a water molecule look like? Does it have a charge?
4. What does hydrophobic and hydrophilic mean?

Materials and Methods

Materials

1. Sand
2. Nanosand
3. Water
4. Cylindrical measurers
5. Containers

Methods

1. Students will be given both regular sand and nanosand.
2. The different sands will be placed in three separate air-tight containers: one containing all regular sand, one containing a 50–50 mixture of sand, and one containing all nanosand. The sand should fill up between ¼ and ½ of the containers and be equal amounts.
3. Students will observe the sands to look for any differences in their dry states.
4. They will then record all observations to compare and contrast later.
5. Purified water will be applied from the cylinder to all three containers. The water should equal the amount of sand used and must also be even.
6. Students will re-observe the sands to search for differences in their wet states.
7. They will again record all observations for comparing and contrasting.
8. Once they have made their recordings, students will have the choice of either using a table or Venn-diagram to illustrate the differences they observed.

Figure 3.7 Nanosand up close. Here a mound of nanosand has been exposed at the water's surface you can see a clear surface tension ring surrounding the dry sand. In addition to a water droplet on the surface of the exposed sand.

Class Questions

1. What do you think is so unique about Nano sand?
2. How is water unable to make the nanosand wet?
3. Why did the regular sand get wet?
4. What are the particles applications of nano coatings?
5. What other materials could we could use to compare the nanosand with instead of regular sand?

References and Suggested Materials

Is Salt Hydrophobic
http://www.answers.com/Q/Is_salt_hydrophobic

Water-repellant sands
http://plantsinaction.science.uq.edu.au/edition1/?q=content/case-study-15-1-water-repellent-sands

Nanoscience Scavenger Card Hunt

Background

The products of nanotechnology are varied and include some common and not some frequently encountered products. Those products fall into a varied of categories: food and drink, medical, environment, textiles, paints (and coatings), transport, computing, energy, cosmetics, and construction. Nano and Me website provides a summary of each type with some examples within the field (http://www.nanoandme.org/nano-products). In this module students will explore Nanoproducts via two activities. One a hands-on exploration of a collection of nanoproducts, students will not only experience hands-on exploration but be able to develop novel application of ideas presented in this activity. Products such as water resistant shirts and pants, antibacterial silver socks, sunscreen, rust-oleum neverwet®, nanocomposite material found in tennis racket or fishing rods, and computer chips and so on. In addition to this exploration students will play a game investigating nanoproducts or not with notecard. After such activities, discussions on nanoparticle technologies may be performed.

Student Learning Objectives

SLO 1 Students will be able to determine the identity most common nanoproducts.

SLO 2 Students should start discussions on how many people use nanomaterials without even knowing it.

Materials and Methods

Materials

1. Various products that use nanoproducts (see background for a list of items).
2. Notecards.
3. Pictures of various nanoproduct and non-nanoproduct items (see Table 3.3).

Methods

Exploring Nanoproducts. Students will be given a container of products and will be allowed to explore by touching, feeling, and discussing potential for that item to use nanotechnology. Proscope micromobile microscopes can be used by students to visualize repellency as water is added to the surface of clothing.

Nanocard Game

Note. This is assuming you want to start from scratch. If already set up, skip to step 5.

1. Teachers will create mini-cards based on the list below.
2. The cards will have a picture on one side and a basic description on the other side.
3. Teachers will also create a dummy card that looks like it should fit with the categories, but either does not exist yet or has nothing to do with nanoscience.
4. The cards will be laid out in random locations or set areas meant to get them to think about how nanoscience would apply to that specific card.
5. Once the student(s) feels that they know where the card goes, they will place it into a designated section category based on Table 3.3.
6. If they feel a card does not fit with anything, they will put that into a junk pile.
7. Whoever gets the most amount of cards correct or, in the event of a possible tie, finds them in the least amount of time can earn a prize (the option will be determined by the teacher).

TABLE 3.3 Summary of Nanoproduct Categories and Products. A Card List

Categories	Card #1	Card #2	Dummy Card Examples
Computing and Electronics	Computer screens	*Quick charge lithium ion batteries*	Touch-sensitive keyboards
Construction	Silica-infused concrete	*Sensors*	Smart tools
Cosmetics	Sunscreen	*Curling tongs*	Makeup
Energy	Batteries	*Building insulations*	
Environment	Smoke filters	*Window glass*	
Food and Drink	Milk	*Ricotta cheese*	Gold-infused water
Medical	Artificial limbs	*Remote drug delivery*	Foreign object detectors
Packaging	Plastic bottles	*Antibacterial coatings*	Water-repellant sealants
Paints	Carbon nanotube paints	*Coatings*	Hydrophobic paintbrushes
Sports and Leisure	Plastic materials	*Ball linings*	Sweat reducers
Textiles and Clothing	Waterproof jackets	*Stain-resistant fabrics*	Quick-dry umbrellas
Transport	Ship hulls	*Smaller and lighter parts*	

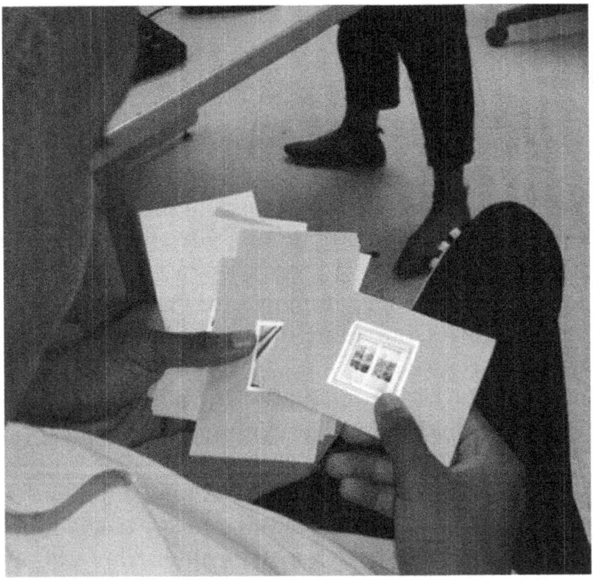

Figure 3.8 Nanoproduct card game. Student is looking through cards to determine if image is in fact a nanoproduct or not. Also what type of potential nonproduct it could be.

Conclusion

Overall, nanoproducts are starting to become a very vital part of our forward progression. Whether they come in the form of water-resistant sealants to protect moisture-sensitive goods, to silver and gold nanoparticles being used as antimicrobials, and nanosand repelling water from important areas, nanoparticles have become very viable in many different fields.

Class Questions

1. What have we about nanoproducts learned today?
2. Are the ideas presented here helpful in real life?
3. Can you think of any other applications for nanotechnologies?

References and Suggested Materials

http://www.photology.info/Hockberger.html%22obi

http://www.webmd.com/beauty/sun/whats-best-sunscreen%22

http://www.badgerbalm.com/s-33-zinc-oxide-sunscreen-nanoparticles.aspx%22

Nano and Me
http://www.nanoandme.org/home/

National Nanotechnology Infrastructure Network. (n.d.). Retrieved July 6, 2015, from http://www.nnin.org/education-training/k-12-teachers/nanotechnology-curriculum-materials

Sondi, I., & Salopek-Sondi, B. (2004, March 18). Silver nanoparticles as antimicrobial agent: A case study on *E. coli* Retrieved July 6, 2015, from https://nanotechnology.wmwikis.net/file/detail/Silver+nanoparticles+as+antimicrobial+article.pdf

Module 4: Silver Nanoparticles As an Antimicrobial?

4

Background

Silver is a naturally occurring metal with germicidal properties that allow it to kill many lower organisms effectively without harming higher animals. For example, silver is capable of rendering stored drinking water potable for a long period of time. Silver ions and silver-based compounds have also been documented to have toxic effects on microorganisms, showing biocidal effects on as many as sixteen different bacterial species, including *E. coli.* As such, silver has been used in dental resin components, ion exchange files, and coatings of medical devices. We are currently testing the effects of silver nanoparticles on basic bacteria to see if they could work as an antibacterial product. This activity will look at the real-life application of silver and gold nanoparticles in a setting that will facilitate critical thinking in a problem-based experiment.

Background for Educators

Nanosilver: Breakthrough for Biohazard
http://www.nisenet.org/catalog/programs/nanosilver_breakthrough_or_biohazard

Student Learning Objectives

SLO 1 Students will be able develop and test hypotheses.

SLO 2 Students will discover real-life applications of silver and gold nanoparticles.

SLO 3 Students will be able to correctly apply size and scale to novel objects.

Pre-Laboratory Questions

1. What do you think silver nanoparticles do?
2. How would the nanoparticles react to normal bacteria?
3. Do you think silver nanoparticles would make a good antibacterial supplement?

Materials and Methods

Safety

- These materials must be handled with care and all dishes must be disposed of in biohazard bags.
- Thermal gloves are recommended while heating agar.

Material

1. Petri dishes
2. Powdered agar
3. Hotplate
4. Graduated cylinder
5. Deionized (DI) water
6. Q-tips
7. Silver-nano particles

8. Meso-silver nanoparticle spray
9. Filter paper
10. Ruler
11. Smart phone
12. Forceps

Methods

1. Students will be given a certain amount of petri dishes.
2. Students will prepare agar by dissolving the appropriate amount of powder into DI water. The mixture will be boiled for one minute, then sterilized in an autoclave. 20 grams of powder to 50 milliliters of water and boil for 1 minute.
3. Fill each petri dish halfway (about 20 mL).
4. Allow the agar to solidify.
5. Once the agar has solidified, bacteria may be sampled then spread evenly on plates.
6. To prepare filter disks, punch holes into 5-filter paper; four disks for one dish.
7. For one sheet of paper, add 0.5 mL of silver nanoparticles; for another, add enough Mesosilver spray to cover it; add 0.5 mL of gold nanoparticles to the last paper sheet; for the control sheet, simply add 0.5 of DI water to the disk.
8. Using forceps, take one disk of each type and place them onto the bacteria dishes in four different positions (see Figure 3.9), then allow to settle for 24 hours.

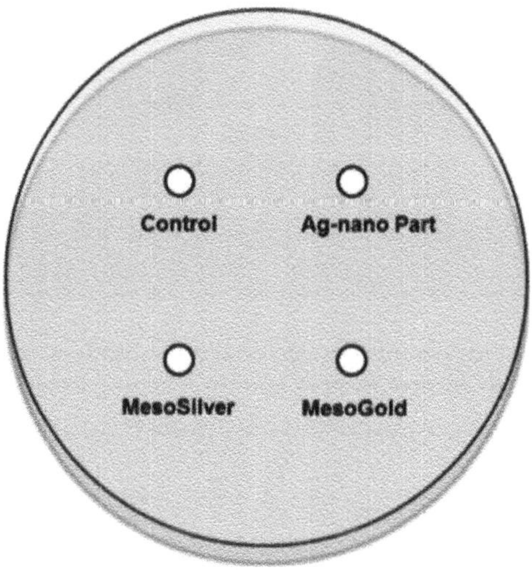

Figure 3.9 Suggested arrangement of treatment disk.

9. Observe dishes and examine for the zone of inhibition (area where the bacteria were unable to grow and the material is exhibiting toxic behavior).
10. To measure the diameter of the zone, measure the zone from one edge directly across to the opposite edge, allowing for a method of qualifying the effectiveness of each treatment (the larger the diameter, the more bacteria have been killed).
11. Record your results with a camera.
12. Have students create a table for each method, report any zones of inhibition, and determine which treatments, if any, acted as an antimicrobial. This can be accomplished by determining means for the zones of inhibition as well as the standard deviation. Ask students to graph these data and present a summary if their hypotheses were supported or rejected.

NOTE: Students maybe be given premade petri dishes with inoculated bacteria such as *Escherichia coli,* and *Staphylococcus aureus* and you can directly start the nanoparticle testing.

Figure 3.10 Student preparation of bacterial trails. Students are treating filter paper disk treated with control (water) and three nanoparticles.

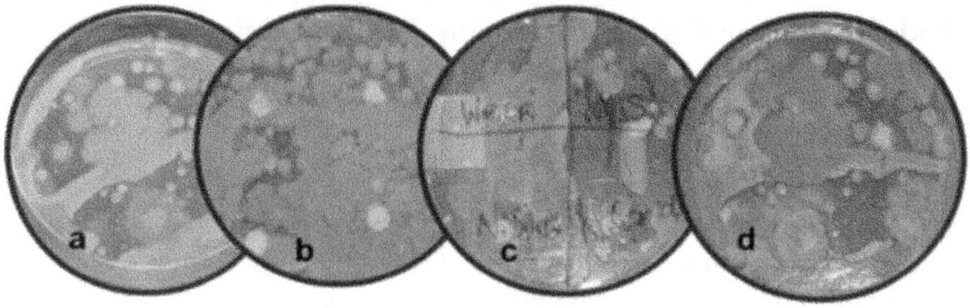

Figure 3.11 Bacterial growth on petri disks at all stages. (a) indicates early bacterial growth, (b) treatment disk are in place, (c) lid labeling to keep track of disk treatment, and (d) results, the disk near d has obvious zone of inhibition.

Class Questions

1. Did adding the nanoparticles make any kind of change?
2. How do the tests compare to the control?
3. What could be observed in the petri dishes from the outcome of these tests?
4. Do you think this would be beneficial to anyone? Why or why not? If so, who?
5. Could this be useful for antibacterial purposes? Why or why not?

References and Suggested Materials

Ammar, R., Al-Warthan, A., Eftaiha, A., & El-Nour, K. A. (2010). *Synthesis and applications of silver nanoparticles.* King Saud University, Retrieved from http://www.sciencedirect.com/science/article/pii/S1878535210000377

Creating and Testing Silver-Nanoparticle Socks. (2012, July 1). Retrieved from http://www.nnin.org/sites/default/files/files/Creating_Silversocks_TG.pdf

National Nanotechnology Infrastructure Network. (n.d.). Retrieved from http://www.nnin.org/education-training/k-12-teachers/nanotechnology-curriculum-materials

Sondi, I., & Salopek-Sondi, B. (2004, March 18). *Silver nanoparticles as antimicrobial agent: A case study on* e. coli as. Retrieved from http://nanotechnology.wmwikis.net/file/view/Silver nanoparticles as antimicrobial article.pdf/279894116/Silver nanoparticles as antimicrobial article.pdf

5

Module 5 Zinc Oxide, Producing Nanoparticles With an Experiment on Their Use

Background

There are a few ways in which nanoparticles may be generated. *Sustainable Nano* has a great introduction at http://sustainable-nano.com/2014/06/10/two-ways-to-make-nanoparticles/. One of the most common nanoparticles produced for educational purposes is zinc oxide.

Zinc oxide can be used in: bandages, pastes, ointments, dental cements, and sunblock. Zinc oxide is used to block the damaging UVA and UVB rays that radiate form the sun. These rays have to ability to damage the DNA in skin cells and can subsequently cause skin cancer (Skin Cancer Foundation, 2015).

The element Zinc has been used for 2,500 years in the production of brass. Metallic Zinc was produced first in India. Zinc oxide, a very popular zinc compound (Philip E. Hockberger Northwestern University, Feinberg School of Medicine) occurs when metallic zinc is exposed to the air and forms a protective coating that protects the rest of the metal. On the other hand, the compound Zinc sulfate glows in the dark. Currently, zinc oxide is used in the production of paint, rubber, cosmetics, and even batteries. All forms of zinc oxide are a sort of white powder that is insoluble in water.

UV Paper

This paper is coated with light-sensitive chemicals (they react to light). This paper will be used to determine if zinc oxide really blocks UV rays. If the paper remains blue after having the cooked zinc oxide nanoparticles spread on it; zinc oxide blocked the UV radiation. However, if the paper turns white; the zinc oxide nanoparticles did not block the UV light.

Ancient civilizations understood the power and importance of the sun. In the 17th century principles began to arise about human relation to the sun. Near the 19th century it was understood that sunlight was not a single stimulus but a collection of stimuli which contains individual (infrared, visible, and ultraviolet). This triggered studies in the area of wavelengths. Thus, leading to the development of a paper that can test for UV rays.

Background for Educators

Nanotechnologies
http://ec.europa.eu/health/scientific_committees/opinions_layman/en/nanotechnologies/l-3/4-nanoparticle-formation.htm

A nanoparticle (or nanopowder or nanocluster or nanocrystal) is a microscopic particle with at least one dimension less than 100 nm
http://www.sciencedaily.com/terms/nanoparticle.htm

Mechanism of Gold Nanoparticle Formation in the Classical Citrate Synthesis Method Derived from Coupled In Situ XANES and SAXS Evaluation
http://pubs.acs.org/doi/pdf/10.1021/ja906506j

Everyday Mysteries, Fun Science Facts from the Library of Congress
How does sunscreen work with a number of sites devoted to dermatology as well as UV radiation
http://www.loc.gov/rr/scitech/mysteries/sunscreen.html

Student Learning Objectives

SLO 1 The student will discuss UV light and its relationship to human health.

SLO 2 The students will explore how nanoparticles are formed and how the formation can be manipulated to produced particles of varying sizes.

SLO 3 The students will demonstrate how zinc oxide blocks UV rays and how the addition to zinc oxide to sunscreen is helpful in skin protection.

SLO 4 The students will generate hypotheses related to UV rays and the harm they can cause to humans and how to prevent such damage.

Pre-Laboratory Questions

1. Why do you think that Zinc Oxide might be used for sunscreen, cosmetics, etc.?
2. Is there a relationship between physical properties and the size of a nano-particle?

Materials and Methods

Materials

1. UV reactive paper
2. Zinc Oxide nanoparticles
3. Sunlight saturated area
4. UV lamps
5. Box and black plastic

Methods

The following methods were copied from University of Wisconsin-Madison MRSEC Education Group (http://education.mrsec.wisc.edu/296.htm)

1. Begin heating a large beaker of water to 65°C.
2. Meanwhile, dissolve 0.10 g Zn(CH3CO2)2 H2O in 25 mL isopropanol with heating in a fume hood. (The actual elapsed time shown in the movie is about 15 minutes.)
3. Meanwhile, prepare an ice bath by adding water to ice. Place 125 mL isopropanol in a flask and chill the flask in the ice bath.
4. When the zinc acetate solid has all dissolved, add that solution to the 125 mL of chilled isopropanol. Also obtain 15 mL of 0.050 M NaOH in isopropanol and chill the solution. (A stock solution may already be chilled. See below.)
5. Slowly transfer the chilled NaOH solution to the chilled and rapidly stirring zinc acetate solution using a pasteur pipet.
6. Place the flask with the mixed solution in the 65°C water bath and start taking samples immediately, taking 1 mL of solution every 30 seconds. Deposit timed samples into 10 mL beakers and allow to cool.
7. Remind students the longer the solution cooks the large the nanoparticles will become. We there for have generated zinc oxide nanoparticles of various sizes. Additionally, that zinc oxide stops the penetrations of UV rays
8. Students should hypotheses how nanoparticle size may affect UV rays absorption.
9. Using a large inverted box with a door cut into the side, and lined with black plastic (trash bag) and with UV lamps suspended from top to create a UV chamber.
10. Using UV reactive paper with grind system (can be drawn on with wax pencil in a low lighted room), paint each timed sample of zinc oxide on the paper, include as well the water (as a control) and include zinc oxide sunscreen as well. This all must be completed out of direct light and in a dim room (as the paper will react).

11. Place this sheet of paper into the UV chamber for 8 minutes removing the paper and treating with dye removal. Examine the resulting grid and determine what size particles prevented a color change induced by UV light.
    (UW MRSEC Education Group, 3115 Engineering Centers Building)

Another method for generating nanoparticles using similar supplies can be found at http://www.education.mrsec.wisc.edu/nanolab/ZnO/index.html

Class Questions

1. How does sunscreen work?
2. Why is sunscreen important?
3. Why would a color change occur when (special) paper is exposed to UV rays?
4. How does zinc oxide particle size affect its ability to block UV rays?
5. If zinc oxide particles are so small they can enter dead skin cells (and there-fore could enter live cells) are there any health concerns?

References and Suggested Materials

Project link: http://education.mrsec.wisc.edu/296.htm

Other links: http://www.photobiology.info/Hockberger.html

A Review on sunscreens and UV light as a health risk. http://www.webmd.com/beauty/sun/whats-best-sunscreen

A sunscreen site providing a context sunscreen http://www.badgerbalm.com/s-33-zinc-oxide-sunscreen-nanoparticles.aspx

Module 6: Nano-Encapsulation

6

Background

Nano-encapsulation is the application of a coating of different substances (known as the shell(s)) around a center (core) material. These shells form a barrier for the encapsulated material and allow for potential applications of materials. We can explore the principle at the macroscale by looking at some common candies such as Malted Balls® and Gobstoppers®.

The process of nano-encapsulation is already commonplace in a range of industries such as in pharmaceuticals for time release medications and targeted drug delivery, pesticide research, and in the food industry. Yet AZoNano (2015) noted while the process is widely accepted only around

10% of potential applications are being exploited. Being familiar with this large field of nanotechnology is a great way to discuss practical applications of nanoscience techniques.

Background for Educators

World's First Nano encapsulated Vitamin is a commercial for an encapsulated product. This video will allow some discussion regarding research and ethics. https://www.youtube.com/watch?v=FqUH0-0y0ql

Student Learning Objectives

SLO 1 Students will be able to understand the process of encapsulation.

SLO 2 Students will have a connection with how the cell membrane functions.

SLO 3 Students will be able to see how encapsulation is incorporated into our everyday lives.

Pre-Laboratory Questions

1. What does encapsulation mean to you?
2. Based on the name, what do you think this experiment will most likely include?
3. In what STEM fields do you think encapsulation would be involved in?

Materials and Methods

Materials

1. sodium alginate (purchased from willpowder.net)
2. calcium chloride (purchased from willpowder.net)
3. liquid laundry starch (purchased from grocery store)
4. bottle of tincture of iodine (purchased at grocery or drug store)
5. Food coloring
6. Water
7. Small disposable pipettes
8. 6 well reaction plate
9. Small plastic spoons or tweezers
10. 5 small cups or beakers

Methods

Modified from NINN. National Nanotechnology Infrastructure Network Outreach Demonstration Guide activity Encapsulation—A New Way for Drug Delivery page 17.

1. To begin place 4 full droppers of calcium chloride solution into the upper left well of the reaction plate
2. In the well below add two full droppers of sodium alginate and 8 to 10 drops of iodine. Allow for the formation of gelatinous spheres. Move promptly to step 3.
3. Draw some of the sodium alginate with iodine mixture into a pipette. Holding the pipette a few centimeters above the surface of the CaCl2 solution begin releasing drops of the alginate-iodine mixture into it. Allow sufficient time for the spheres to develop.
4. Place into the next lower well of the reaction plate liquid laundry starch solution (starch mixed with water). Take a few of the spheres that are created in well one and place in the starch solution well (Figure 3.12).
5. Return to well two and draw some of the sodium alginate with starch mixture into a pipette. Holding the pipette a few centimeters above the surface of the CaCl2 well begin releasing drops of the alginate-starch mixture into it.
6. Place into the upper right well of the reaction plate iodine water (water that a few drops of iodine have been added to). Take the spheres that are created in well one and place in the iodine solution well.

Figure 3.12 Set-up of encapsulation. a) preparations of wells and b) encapsulated sphere

7. Have students look at the changes that occur in the spheres that were placed in the starch water and the iodine water. The spheres that are in the middle well should show that the starch water is getting blue or black because the iodine can travel through the outer covering of the sphere. The spheres that are in the iodine water should show that the spheres are turning blue or black because the starch molecules are too big to travel through the sphere outer covering but the iodine molecules are small enough to travel into the spheres through the covering.

Explain to students that observing the combinations as they appear will provide insights into the molecular processes (nanoscale) that go on inside and outside of the macrospheres. This will introduce them to an important biological phenomenon that occurs in cells and will be the first step towards understanding how encapsulation can be used in the battle against a disease like cancer.

You can also do this activity with Gaviscon which is a commercial antacid that contains sodium alginate as a thickening agent.

Extended Activity

As noted previously encapsulation is greatly relied on in the pharmaceuticals fields. Review the following YouTube video as one example of time release pills (https://www.youtube.com/watch?v=uojwMhQpjq8).

Review some of the basic steps in digestion where pH changes, where and what enzymes enter the system. If you were to design a drug that would suppress appetite for an entire day. Detail how you would design this drug and trace its release through the digestive system. Use Figure 3.13 to help you.

Class Questions

1. Based on what you just learned: what is encapsulation?
2. How do you think encapsulation would help everyday people?
3. How much fun did you have? (Scale from 1–10)

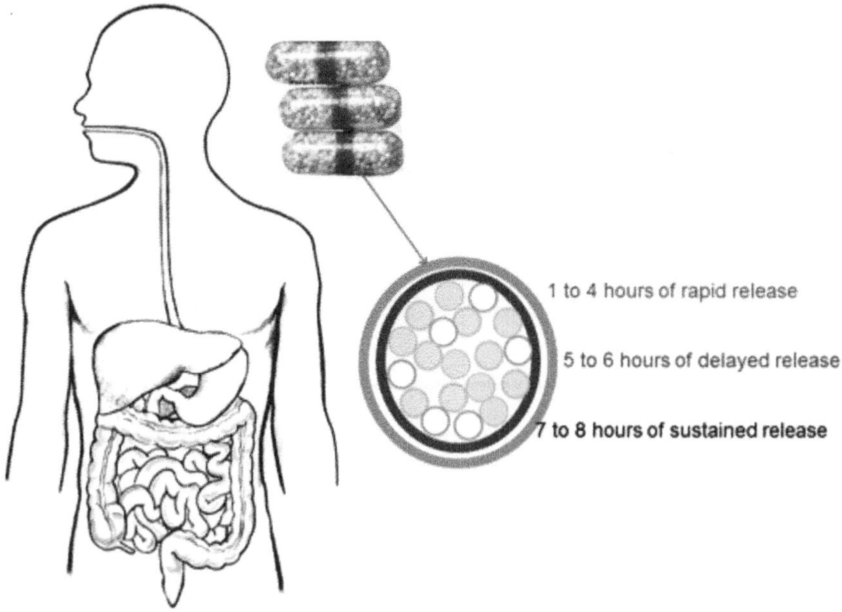

Figure 3.13 Diagram of Alimentary Canal and cross section of slow release medicine. Students should be able to apply anatomy as a guide for medication release.

References and Suggested Materials

National Nanotechnology Infrastructure Network Outreach Demonstration Guide
http://www.nnin.org/sites/default/files/files/NNIN_Outreach_Demo_Guide-11_10_0.pdf

Nano encapsulation of food ingredients using lipid based delivery systems
http://www.excli.de/vol13/Yadav_20032014_proof.pdf

AZONano
http://www.azonano.com/article.aspx?ArticleID=1875

Module 7: NanoScience in Nature: Build a Beetle

7

Background

Nanotechnology has been in nature for a long time. One example are unique bright red sunsets which are caused by nanoparticles introduced into our atmosphere from volcanic eruptions. Many different animals and plants also have unique properties because of their nanostructure. The beautiful wings of Blue Morpho Butterflies (*Morpho peleides*) [http://www.rainforest-alliance

Figure 3.14 *Stenocara gracilipes, Fogstand Beetle* or Namib Desert Darkling Beetle. Using a nanoscturure on the forewings to collect and condensate water from fog.

.org/kids/species-profiles/blue-butterfly] consist of nanostructure that cause them to appear multicolored. A Gecko can walk upside down on many types of surfaces. This is because the lizard's feet are covered with nanospatulae hairs which allow this unique ability based on close molecular attraction. As a result of the evolution of nanoscience we are now able to mimic many adaptations of animals and plants for the benefit of humankind such as beetle inspired water retrieval systems (see Figure 3.14). Another example is the lotus effect, which is used to describe the fact that in nature a lotus plant is hydrophobic which allows it to repel water. With Nanotechnology we are able to find ways to make materials that mimic this hydrophobic property displayed by lotus plants. The possibilities of products that can be developed using nanotechnology are endless.

Background for Educators

Hydrophobic Materials

https://www.youtube.com/watch?v=z4EFycryEU4

Geckos' Sticky Secret? They Hang by Toe Hairs

http://www.livescience.com/47307-how-geckos-stick-and-unstick-feet.html

Student Learning Objectives

SLO 1 Students will draw connection between nature and nanoscience.

SLO 2 Students will understand the concepts of how we can use nanoparticles for our advantages.

Pre-Lesson

The students should watch the following video: Video explaining and showing Beetles in the Sahara desert. https://www.youtube.com/watch?v=XsMJNNshPOs

Pre-Laboratory Questions

1. Do nanoparticles exist naturally?
2. Where would you find nanoparticles in nature?
3. What are some of the influenced on the natural selection in nature? Would such selective factors lead to nano-solutions?
4. Do you think there could be biological-inspired nanotechnology?

Materials and Methods

Materials

1. Construction paper
2. Scissors
3. Silhouette of beetle

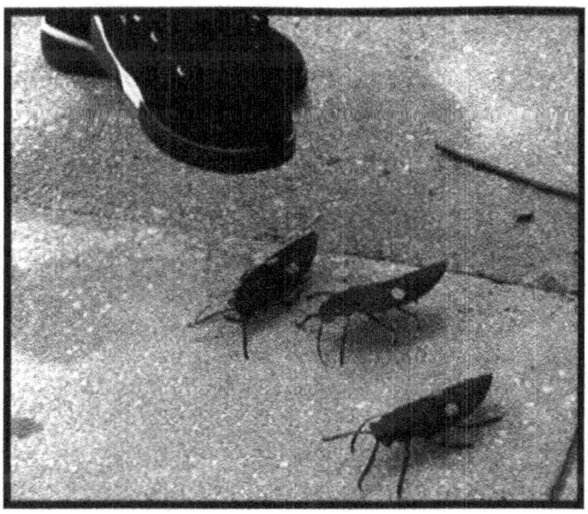

Figure 3.15 Student Mock-Desert Beetles. Beetles ready for fog treatment.

4. Spray bottle
5. Water
6. Rust-Oleum NeverWet

Methods

Use the silhouette of the beetle provided as a template to cut out individual beetles.

Beetle Template

1. Cut out the beetle silhouette on dark (allows water droplets to be seen easier) construction paper (The number of the Beetles cut out depends on class size).
2. Go outside or well-ventilated space to treat the paper beetles.
3. Wearing gloves and eye protection, lay each beetle down on a flat surface and apply a coating of Rust-Oleum NeverWet®; use a paintbrush to evenly coat the Rust-Oleum on beetles).

4. Let the beetles dry for 5–10 min. Then apply another coat. In total aim for 2–3 coats.
5. Ask Students to hypothesize what will happen when the treated beetles are exposed to water.
6. Spray water onto the beetles to simulate the morning fog (Figure 3.16; Pre-lesson video).

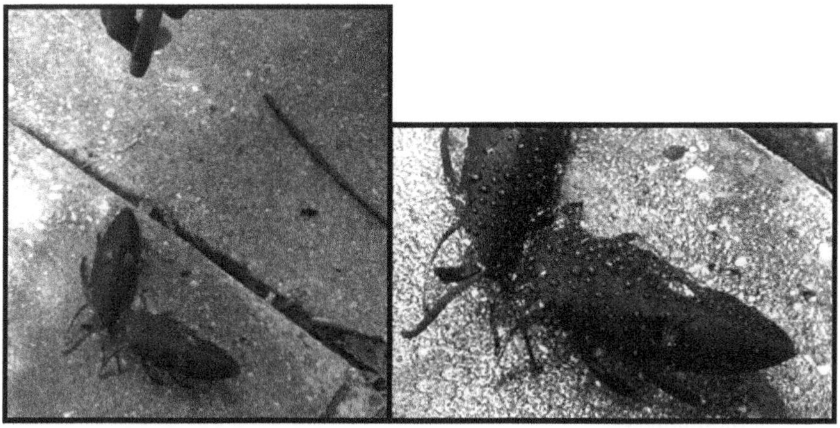

(a) (b)

Figure 3.16 Beetles and Hydrophobicity. (a) Student beetles being treated with (b) mist to test their hypotheses.

7. Have a discussion on what the students have just observed; could such models in nature be used to solve real world problems?
8. After discussion have students watch. The following TED talk (or pre-assign).

   Janine Benyus: Biomimicry in Action. Janine Benyus has a message for inventors: When solving a design problem, look to nature first. There you'll find inspired designs for making things waterproof, aerodynamic, solar-powered and more. Here she reveals dozens of new products that take their cue from nature with spectacular results. https://www.ted.com/talks/janine_benyus_biomimicry_in_action

Biomimicry is an approach to innovation that seeks sustainable solutions to human challenges by emulating nature's time-tested patterns and strategies. The goal is to create products, processes, and policies—new ways of living—that are well-adapted to life on earth. A summary of the field of Biomimicry can be found at http://biomimicry.net/about/biomimicry/a-biomimicry-primer/. The Fogstand Beetle *Stenocara gracilipes* for example has been used as a model to retrieve water from air. Merging fields of nanoscience and biomimicry allows for the production of drinkable water and water for use of agriculture.

Self-Filling Water Bottle Takes Cues From Desert Beetle

http://phys.org/news/2012-11-self-filling-bottle-cues-beetle.html

Beetle's Shell Offers Clues to Harvesting Water in the Desert

http://news.nationalgeographic.com/news/2001/11/1101_TVdesertbeetle.html

Facilitate another discussion referencing their beetles, brain storming some problems the world faces and how bio-inspired nanotechnology can solve them.

Class Questions

1. Why would the beetle evolve to be hydrophobic?
2. Do you think there is other desert animals that us this strategy?
3. What is the biomimicry?
4. Is there a relationship between nanotechnology and the field of biomimicry?
5. What other fields or aspect of nanoscience can be linked with biology?

References and Suggested Materials

Biomimicry—The Namib Desert beetle a source of inspiration
http://morawatersystems.com/biomimicry-the-namib-desert-beetle-a-source-of-inspiration/

4

In the Classroom

As you set up your elementary or secondary science classroom, make sure you allow room for mistakes and wonder. What does this mean? So often in today's classrooms, teachers are terrified of making mistakes, or of their students making mistakes... with the school year clock ticking and the standardized tests looming. Science *is* about mistakes. Most of the scientists working in labs today gain more information from experiments when things don't go as planned, than they do when they come out as expected. In fact, in science, we don't describe the world as one where there are successes and failures; but rather, attempts and outcomes. When students are taught about attempts and outcomes, it is amazing how their attitudes change. They become more willing to take risks, to experiment and to follow paths that they wouldn't have previously. But teachers must model this behavior. Teachers who show how real science is done, become role models for students to pursue STEM fields.

Nanoscience Research Modules for Pre-Service STEM Teachers, pages 57–60

Lesson Planning

This book is set up to present these nanoscience lessons as modules, however thanks to Bloom's Taxonomy, all teachers can fit these modules into lesson plans specific to their needs (see Figure 4.1). Benjamin Bloom was an educational psychologist who divided the levels of learning into a hierarchy, from lowest to highest. At each level of learning there are corresponding verbs, making it easy to write daily objectives for lesson plans. For example; a knowledge-based daily objective would be: "The student will be able to list the parts of a plant cell," with the verb *list* as a knowledge level verb. Daily objectives are crucial in writing lesson plans, as it gives the actual goals you are teaching towards. Indeed, you can say that you are "teaching to the objective" when you write good objectives.

Teaching to the Objective

Would you want to plan a car trip across the country without an end goal or objective? The same can be said of a well written lesson plan. You write a good objective (or objectives) with Bloom's taxonomy verbs, you teach in order to reach that goal, and then you measure or assess the students

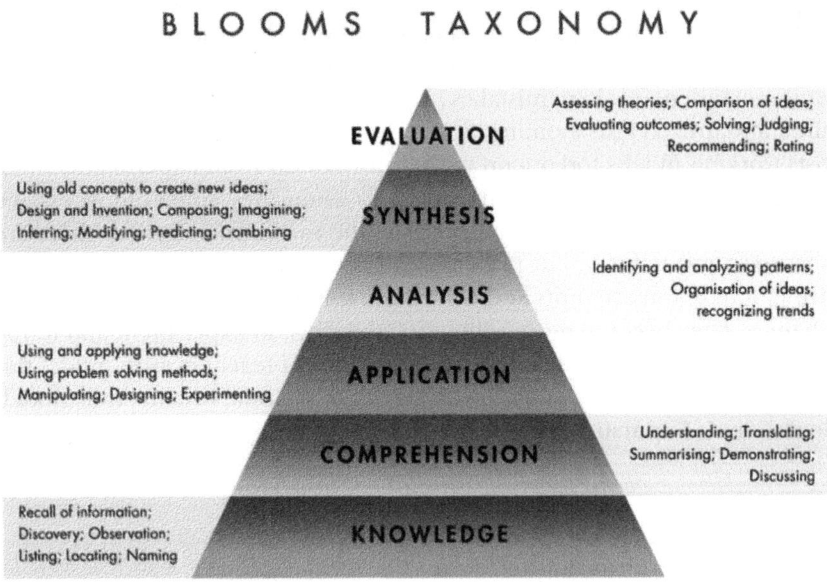

Figure 4.1 Bloom's taxonomy. *Source:* Cornwell (2011), https://juliaec.wordpress.com/2011/03/23/blooms-taxonomy-encouraging-higher-cognitive-thinking-in-primary-school-classrooms/

on that objective. This is the heart and soul of all good lesson plans. "The student will be able to".... with this beginning statement of your daily objective, you are setting the final goal as the first and most important aim of your lesson. You are not leaving it up to chance; you are saying that because of your teaching, the students will be able to "do" something that they weren't able to do before. Not only that, but the thing that they are able to "do," is measurable and observable. It is a behavior. It's not enough for students to tell you that they "understand" something. How do you know that they understand?

Many state-wide standards are written with broad language, and are not meant to tell teachers "how" they go about teaching the content. For example; the Virginia Standards of Learning (SOLs) are writing in very general terms, meant to relay the content that will be taught during a school year. For 7th grade life science, one of the SOLs states;

> LS.2 The student will investigate and understand that all living things are composed of cells. Key concepts include
> (a) cell structure and organelles;
> (b) similarities and differences between plant and animal cells;
> (c) development of cell theory; and
> (d) cell division (Virginia SOL Website: http://www.doe.virginia.gov/testing/sol/standards_docs/science/index.shtml).

The word *understand* is mentioned but does not describe how the teacher will know that the student will understand. This is where the daily objective comes in. Daily objectives, as mentioned before, are written near the beginning of every lesson plan and describes things that the students will be able to do as a result of the lesson, using Bloom's taxonomy levels of learning and the corresponding verbs. So again, if the daily objective states: *The student will be able to list the parts of the plant cell,* then the teacher is teaching at the knowledge level of Bloom's taxonomy because the verb *list* is mentioned in that level. The teacher will teach so that the student will be able to provide her with a list of the parts of the plant cell as an assessment for that lesson. In this way, the teacher is "teaching to the objective;" in other words, the teacher is making sure that the objective is met by measuring the objective at the end of the lesson. This ensures a well-aligned lesson plan; one in which the daily objective, the instruction and the assessment, are all saying the same thing. It is the same as plotting out your end result as a marathon runner, with the end in mind (crossing the finish line) as the first step. The course is subsequently charted so that the end goal will be met. If you don't cross the finish line, you have not met your objective.

When writing STEM lesson plans, many times they will be integrated and will include math and science content in the same lesson. It is ok to have more than one daily objective, and it is ok to take more than one day to achieve these goals. As stated earlier, make some space for "mistakes" or attempts and outcomes in your science classroom. This is where the real science is being learned. Many "aha" moments occur only when something does not go as planned.

Teaching is an art, a profession, and a craft. It takes time to be able to write perfect lesson plans. Part of that perfection however, is realizing that they can change at any time and that a teachable moment may arise and in effect re-write your lesson as you are teaching it. That's ok. It's more important to teach your children than to bore them with test preparation and lower level teaching for the whole school year. It is the hope of the authors of this book, that elementary and secondary teachers will find some confidence in teaching nanoscience to students using Bloom's taxonomy with well written lessons. Since nanoscience is a cutting-edge field, it is to be expected that with new technologies there will be new discoveries and breaking news on a regular basis. The wonderful thing about science is that it is different from other content areas in this way, and it takes courageous and adventurous teachers to bring children along on the ride. Use these modules and make them your own, no matter the grade level. Your most important job is to create scientifically literate students who can face the world with the ability to confidently make decisions that will not only affect their lives, but the lives of future generations to come. It is a most important job and we know you can do it.

References

Bell, R., Smetana, N., & Binns, I. (2005). Simplifying inquiry instruction. *The Science Teacher, 72*(7): 30–34.

Berube, C. (2014). *STEM and the city; A report on STEM education in the great American urban public school system.* Charlotte, NC: Information Age Publishing.

Bowles, K. (2004). Teaching nanotechnology in the high school curriculum: A teacher's guide. *Apopka High School; UFC NANOPAC, 1,* 1–43.

Carter, A. (2001). Interactive distance education: Implications for the adult learner. *International Journal of Instructional Media, 28,* 249–161.

Chauhan, S., & Taylor, P. (2013). Access & use of science and technology in the classroom: Ensuring academic success through STEM preparedness. Retrieved from http://www.multiculturaleducationalliance.org/wp-content/uploads/2013/06/access_adoption_use_of_science_and_technology.pdf

Chang, R. P. H. (2006) A call for nanoscience education. *Nanotoday, 1*(2), 1–7.

Clewell, B., & Campbell, P. (2007). *Good schools in poor neighborhoods.* Washington, DC: The Urban Institute Press.

Cornwell, J. (2011). *Bloom's taxonomy: Encouraging higher cognitive thinking in primary school classrooms. Successful Teaching Blog,* March 23. https://juliaec.wordpress.com/2011/03/23/blooms-taxonomy-encouraging-higher-cognitive-thinking-in-primary-school-classrooms/

Drake, K. N., & Long, D. (2009). Rebecca's in the dark: A comparative study of problem-based learning and direct instruction/experiential learning in two 4th grade classrooms. *Journal of Elementary Science Education, 21*(1), 1–16.

Nanoscience Research Modules for Pre-Service STEM Teachers, pages 61–63

Etherington, M. (2003). Investigative primary science: A problem-based learning approach. *Australian Journal of Teacher Education, 36*(9), 36–57.

Filipponi, L., Sutherland, D., & Center, I. N. (2010). Introduction to nanoscience and nanotechnologies. *NANOYOU Teachers Training Kit in Nanoscience and Nanotechnologies,* 1–29.

Gonzales, P., Williams, T., Jocelyn, L., Roey, S., Kastberg, D., & Brenwald, S. (2008). *Highlights from TIMMS 2007: Mathematics and science achievement of U.S. fourth- and eighth-grade students in an international context.* Washington, DC: National Center for Education Statistics, Institute of Education Sciences, U.S. Department of Education.

Haberman, M. (1995). *Star teachers of children in poverty.* West Lafayette, IN: Kappa Delta Pi.

Haberman, M. (2005). *Star teachers: The ideology and best practice of effective teachers of diverse children and youth in poverty.* The Haberman Educational Foundation.

Haycock, K. (2001). Closing the achievement gap. *Educational Leadership, 58*(6), 6–11.

Krajcik, J., Blumenfeld, P. C., Marx, R. W., Bass, K. M., Fredricks, J., & Soloway, E. (1998). Inquiry in project-based science classrooms: Initial attempts by middle school students. *Journal of Learning Science, 7,* 313–350.

Light, G., Swarat, S., Park, E. J., Drane, D., Tevaarwerk, E., & Mason, T. (2007). Understanding undergraduate students' conceptions of a core nanoscience concept: Size and scale. In *Proceedings of the First International Conference on Research in Engineering Education* (pp. 1–13). Honolulu, HI.

Michael, J. (2006). Where's the evidence that active learning works? *Advances in physiology education, 30*(4), 159–167.

National Math and Science Initiative; Transforming Math and Science Education. (2015). White Paper: *Increasing the Achievement and Presence of Under-Represented Minorities in STEM Fields.* nms.org, Dallas, TX: nms.org.

Nicholls, G. M., Wolfe, H., Besterfield-Sacre, M., Shuman, L. J., & Larpkiattaworn, S. (2007). A method for identifying variables for predicting STEM enrollment. *Journal of Engineering Education, 96*(1), 33–44.

Peri, M., Grigg, W., & Dion, G. (2005). *The nation's report card: Mathematics 2005.* NCES 2006-453. Washington DC: U.S. Department of Education, Institute of Educational Sciences, National Center for Educational Statistics.

Price, M. (2012). Pushing students toward STEM. *Science Career Magazine.* Retrieved from http://sciencecareers.sciencemag.org/career_magazine/previous_issues/articles/2012_07_06/caredit.a1200076

Pollard-Durodola, S. (2003). Wesley Elementary: A beacon of hope for at-risk students. *Education and Urban Society, 36*(1), 94–117.

Reeves, J. (2006). Teachers as facilitators: What autonomy-supportive teachers do and why their students benefit. *The Elementary School Journal, 106*(3), 225–236.

Rice, J. K. (2003). *Teacher quality: Understanding the effectiveness of teacher attributes.* Washington DC: Economic Policy Institute.

Skinner, E., & Chi, U. (2012). Intrinsic motivation and engagement as "active ingredients" in garden-based education: Examining models and measures derived from self-determination theory. *The Journal of Environmental Education, 43*(1), 16–36.

Sohlberg, K. (2006). Introducing the core concepts of nanoscience and nanotechnology: Two vignettes. *Journal of Chemical Education 83*(10), 1516.

Sondi, I., & Salopek-Sondi, B. (2004). Silver nanoparticles as antimicrobial agent: A case study on *E. coli* as. Retrieved from http://nanotechnology.wmwikis.net/file/view/Silver nanoparticles as antimicrobial article.pdf/279894116/Silver nanoparticles as antimicrobial article.pdf

Smith, T. R., McGowan, J., Allen, A. R., Johnson, W. D., Dickson Jr., L. A., Najee-ullah, M. A., & Peters, M. (2008). Evaluating the impact of a faculty learning community on STEM teaching and learning. *The Journal of Negro Education,* 203–226.

Summers, M. F., & Hrabowski, F. A. (2006). Preparing minority scientists and engineers. *Science 31,* 1870–1871.

Thomas, K., Horne, P., Donnelly, S., & Berube, C. (2013). Infusing problem-based learning (PBL) into science methods courses across Virginia. *The Journal of Math and Science: Collaborative Explorations, 13,* 93–110.

Virginia Standards of Learning. (2010). Retrieved from http://www.doe.virginia.gov/testing/sol/standards_docs/science/index.shtml

VISTA: Virginia Initiative for Science Teaching and Achievement. (2015). George Mason University. Retrieved from http://vista.gmu.edu/

Ward, A. R. (2008). *Content knowledge and classroom inquiry style: Factors influencing inquiry-based science teaching practice of elementary student teachers* (Doctoral Dissertation). Retrieved from ProQuest Dissertations and Theses. UMI number 3325997.

Wood, L., & Gebhardt, P. (2013). Bioinformatics goes to school—new avenues for teaching contemporary biology. *PLoS Computational Biology, 9*(6).

About the Authors

Shawn T. Dash, BSc, MSc, and PhD. Assistant Professor in Biological Sciences in the School of Science at Hampton University. Principal Investigator and Mentor for NanoHU Pioneers Summer Research Program. Teaching a variety of courses including introductory and organismal biology, ecology, entomology, invertebrate zoology, desert ecology, zoology, medical entomology, animal behavior, birds and mammals, anatomy and physiology, animal ecology, and human biology, Dr. Dash strives to share his passion for biology. Frequently an invited speaker to local nature organization. Dr. Dash's research areas focus on biodiversity and systematics of invertebrates particularly ants. In addition to his myrmecological research Dr. Dash is dedicated and focused on science education. With a deep interest in using taxonomy as a means to introducing the scientific process to undergraduates.

Dr. Clair Berube is an Associate Professor of Science Education at Hampton University in Hampton, Virginia. Before coming to Hampton, she served as Visiting Assistant Professor of Science Education at the College of Staten Island, City University of New York. She taught science education courses and graduate science education research courses. Dr. Berube was also a fellow at the Discovery Institute, an education think tank at the College of Staten Island. Dr. Berube taught science education at Wagner College, New York City and at Old Dominion University, Norfolk, Virginia. From 1995–1997 Dr. Berube was a science teacher, (6th & 7th Grade) St. Gregory the Great Catholic School, Virginia Beach, Virginia. From 1992–1995 Dr. Berube served as a science teacher, (6th & 7th Grade) at Blair Middle

Nanoscience Research Modules for Pre-Service STEM Teachers, pages 65–67

School, Norfolk Public Schools. Duties included co-director of CHROME after school program for girls/minorities in science and technology. While at Blair Dr. Berube won a teaching award teaching science. Dr. Berube serves on various professional committees, including Virginia Association of Science Teachers and VAST College Committee, (VSEP) Virginia Science Education Professors, VMSC (Virginia Math/Science Coalition), and VSELA (Virginia Science Education Leadership Association). She is a member of American Educational Research Association SIG Science Teaching and Learning. Dr. Berube has made several presentations at regional, national and international conferences. She has also served as a peer-reviewer for national STEM education journals. Her latest book is *STEM and the City; a report on STEM Education in the Great American Urban Public School System,* (2014), Information Age.

Dr. Cindy Thomas-Charles is currently an Assisted Professor of Biological Sciences at Hampton University. She currently also serves as Director of the National Institute on Health (NIH)-funded Hampton University Maximizing Access to Research Careeres Program, co-investigator of the Hampton University Robert Noyce Teacher Scholarship Program, and as coordinator of the biology masters program. Dr. Thomas-Charles completed a bachelors degree in biology, and doctor of philosophy degree in genetics from Stony Brook University in Stony Brook, NY. During her graduate training she was the recipient of a National Science Foundation-funded Bridge to the Doctorate Fellowship. Her dissertation research was also supported in part by the NIH-funded Molecular and Cell Biology of Infectious Disease Training Program, and the Dr. W. Burghardt Turner Dissertation Fellowship.

NanoHU Pioneers Summer Research Program of 2015

Kevon E. Barrett, current member of NanoHU Pioneers Summer Research Program of 2015. I'm a rising 10th grader attending Phoebus High School, in Hampton, Virginia. I have a strong passion for STEM. The developments of STEM opportunities are increasing and I want to be a part of them. Nanoscience makes a great connection with STEM, making STEM more understandable. With the knowledge of STEM I know I can do big things. My future plans are to attend college and get a degree. I currently don't have a specific field I would like to major in, but by attending this program they will lead me into the right direction. In this program I have seen that forensic is interesting and it is something I would like to look into. "Do the habits you have today, good enough for the goals you have tomorrow."

Sydney Alexandria Hardy I attend the Governors STEM Academy at Grassfield High School, located in Chesapeake, Virginia. I am a 10th grader interested in STEM and Nanoscience: I love STEM because it is fun and exciting. The development of STEM/NANO has helped medicine and other useful areas of life. Upon graduating High School, I plan to attend a 4-year university. I would either like to go into Biomedicine or Law. I am between the two and I have not made a final decision (one will be made soon) However, my strong STEM background will help me in whatever I choose. My dream is to earn a PhD in the discipline of my final decision. I love education and I strive to be the best.

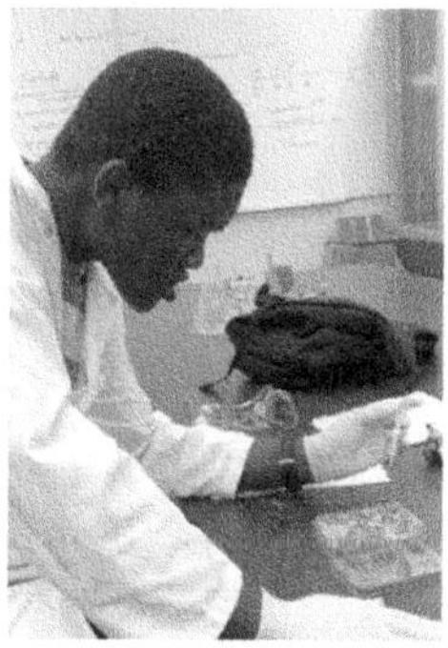

Tomar M. Pettus Jr. I attend Tabb High School, in Yorktown Virginia. I am a rising 11th grader extremely interested in STEM which is something that has always appealed to me. I am interested in nanoscience to a moderate degree, but I am deeply interested in computer engineering. After High School I hope to attend the Massachusetts Institute of Technology majoring in Computer Engineering.

Printed and bound by CPI Group (UK) Ltd, Croydon, CR0 4YY

15/09/2026

14962335-0006